INTERNET & MOBILE ABUSE

HOW TO OVERCOME?

DR. P.N. SURESH KUMAR

INDIA • SINGAPORE • MALAYSIA

Notion Press Media Pvt Ltd

No. 50, Chettiyar Agaram Main Road,
Vanagaram, Chennai, Tamil Nadu – 600 095

First Published by Notion Press 2021
Copyright © Dr. P.N. Suresh Kumar 2021
All Rights Reserved.

ISBN 978-1-63904-507-5

This book has been published with all efforts taken to make the material error-free after the consent of the author. However, the author and the publisher do not assume and hereby disclaim any liability to any party for any loss, damage, or disruption caused by errors or omissions, whether such errors or omissions result from negligence, accident, or any other cause.

While every effort has been made to avoid any mistake or omission, this publication is being sold on the condition and understanding that neither the author nor the publishers or printers would be liable in any manner to any person by reason of any mistake or omission in this publication or for any action taken or omitted to be taken or advice rendered or accepted on the basis of this work. For any defect in printing or binding the publishers will be liable only to replace the defective copy by another copy of this work then available.

Contents

Preface *5*
Translation Notes *7*
Prologue *9*

1. The Spreading Disease 11
2. Who Becomes Slaves of the Internet? 17
3. Internet Addiction – The Effect on the Body 21
4. The Online Activities of College Students 25
5. Internet and Occupation 31
6. How to Realize Internet Addiction? 35
7. Cyber Sex Addiction 43
8. The Cyber Threat – Problem of the Future Generation 53
9. Cyber Crime 57
10. Prevention and Treatment 63
11. Mobile Phone Addiction 73
12. Misuse of Mobile Phones 77
13. Computer Game Addiction 81
14. What is Blue Whale Game? 87
15. Selfitis – The New Disease of the New Age 91
16. The Use of Mobiles and Health Problems 99
17. Treatment 107

Preface

The world-wide misuse of mobile phone and computers have inversely affected our relationships, friendships and responsibilities towards our jobs. It has created a new generation which is addicted to these electronic equipments. This tendency has resulted in unhealthy conditions which affect the body and the mind.

The tendency to misuse electronic equipment is a world-wide phenomenon. Asian nations have declared this as a national health problem. This has widely spread to the first world countries such as North America and Europe.

Internet addiction becomes uncontrollable as it affects both body and mind. We become highly involved with our online friends emotionally. Some sit for hours together before the computer looking for subjects of interest or blogging. Some want to know the day to day affairs of the society and collect the opinion of other people on these subjects.

Those who cannot connect with people become slaves of the internet. This book discusses about the tragic state of the new generation and tries to persuade them to outlive their addiction and to persuade them to seek medical help.

Dr. P.N. Suresh Kumar, MD, DPM,
DNB, PhD, MRC Psych
Professor of Psychiatry
KMCT Medical College
Kozhikode, Kerala, India

Translation Notes

This is a token of gratitude and admiration to Dr. P.N. Suresh Kumar who has dedicated his life for mental health of the society. I have watched him grow from better to best among the psychiatrists. He is always up to date in his field of study. I wonder where he finds time to write books and present papers. A simple passion by nature, his simplicity is the language in his books. They are also very scientific in methods of treatment.

As a parent, this book will help you to be aware of the behavioral changes in your ward. Even if you have a "perfect" son or daughter, you can be aware of the pitfalls of teenagers in general. Interests may grow into addictions and deep addictions any time. Beware!

I proudly present the book for the young and old of modern times. I volunteered to translate the content from Malayalam to English because internet addiction is rampant in higher strata of teenagers who are more familiar with English. I have PhD in Malayalam language and literature and was the head of department of Providence Women's College, Calicut. I like to take up the challenge of translating books from English to Malayalam and vice versa

Dr. Nalini Satheesh, MA, PhD
Former Professor & Head
Department of Malayalam Language
Providence College
Kozhikode, Kerala, India

Prologue

This book discusses the problem of internet addiction through the insight of a psychiatrist and suggests means to stop addiction. This is one of the books among several which deal the same subject. The simplicity of language and presentation is unique and provides the reader with a lot of knowledge and satisfaction.

The book is an eye opener to the new disease of internet addiction which not only affects the body but also the mind. It disrupts the peace of everyday life. The problem is very serious because the addicts are below thirty years of age. Dr. P.N. Suresh Kumar warns the decay of natural relationships which will destroy the social life altogether. He persuades us to be aware of this grave situation.

How can we safeguard the new generation from the hands of the octopus which become stronger by the day? Internet and mobile have become an integral part of our lives. The female section of the society is trapped in strange relationships developed through face book and WhatSapp. This book reminds us to be aware of the situation which is serious. The internet is surely the implement of modernity. But how we can avoid the bad effects of these equipments which cause harm to body and mind is the subject discussed in the book. We present you with this rare gift of awareness with pride and satisfaction.

Indian Medical Association
Mental Health Committee, Kerala State

Chapter 1

The Spreading Disease

Dr. Kimberly, the famous American doctor was the first person to study about the internet addiction. He presented a paper at the venue of American Psychological Society during its annual meeting in 1996. This paper was named – 'Internet Addiction – the emerging of a new Disorder'. The disorder quickly spread to other nations such as Australia, China, Korea and Taiwan. This has now affected the whole world.

The Chinese Internet Network declares that 513 million Chinese citizens are slaves of the internet. This comes to 55.4% of Asians and 23.2% of people in the world who are addicted. The China Communist League declared in 2007 that 17% of children  between 13 and 17 have become slaves of the internet. Nihan University of Japan has found out that 8% of one lakh school children have become internet addicts.

When people sit before the computer it is difficult to find out whether they are using the internet or using it for some project. Computer use affects their mind and emotions and they lose their individual personality. Some may become prey to anxiety disorders.

Mareca Oreak of Harvard University which studies computer generated disorders declare that 5-10% of regular internet users are slaves of the computer. The Centre for Internet Addiction Recovery, headed by Dr. Kimberly says that acute depression patients immerse themselves in the magic world of internet to escape from reality. 60% of the addicts see pornographic picture or use pornographic language for pleasure. Some of them are also addicted to toxic liquor and narcotic drugs.

Reasons and Result

Dr. Yung, the famous psychiatrist emphasizes the strong influence of internet addiction in the social psychological and occupational fields. Some use the internet for educational purposes. Some need it for their occupation. But those who use the internet for more than 30 hours a week are worse affected in educational standards. The bonds of marriage suffer. The efficiency decreases in the field of occupation. The study done in Korea supports Dr. Yung's opinion.

Dr. Yung has found out in his study that 52% of internet slaves use liquor, narcotic drugs and many have gambling addiction. They use internet as a means to escape from the pressure of life. May be the internet addiction is not harmful to the extent of liquor addiction, but it will certainly affect your mind and body in an adverse way.

The Negative Effects

Just as liquor, narcotic drugs, internet addiction affect the family life, financial security, efficiency in job, and educational standards. The internet addicts move away from personal relationships, family bonds and build a magical world of their own. They will try to hide the fact that they are spending a lot of time on the internet. Some try to hide their personality behind some other name and pretend to be of a character built by their own imagination. They do this because of inferiority complex or because they do not get enough appreciation from others in real life. These lead them to depression and anxiety neurosis.

Those who are trying to get out of addiction may experience anger, sorrow, fear, restlessness. They may also complain of stomach ache, pain of finger tips (Carpal Tunnel Syndrome) or severe headache. They may have sleep disorder. The addicts often lead a disorderly life unmindful of personal hygiene.

Symptoms

Are you an addict of the internet? If you have given the answer 'yes' to 5 of these 8 questions, you should seek the help of a psychiatrist.

1. Do you always think of the internet for doing something or other?
2. Do you want to extend the time on internet to gain more mental satisfaction?
3. Has your ability to get away from the internet or control your habit has failed?
4. Are you exceeding the pre-planned time when using the internet?
5. Trapped in the internet have you failed in relationships, educational endevours and job excellence?
6. Are you trying to hide your addiction from family members and psychiatrists?
7. Do you feel hopelessness, guilt, anxiety and depression and escape into the world of the internet?

Other Symptoms

1. Failed attempts to control addition
2. The pleasure one feels when using internet
3. Ignoring friends and relatives

4. Hiding many things from others
5. Feeling guilty, anxious and sadness because of the addiction
6. Weight loss or obesity, back pain, headache and pain in the hands
7. Moving away from other pleasurable activities.

Internet addiction is a way to escape the pressure of life. The addicts also resort to seek pleasure in an unreal world. There are three types of addiction. Video game addiction, cyber sex addiction and online gambling addiction. Cell phone addiction, smart phone addiction, and face book addiction are also related to the above. The addictions are inter-related. Those who have made a study of internet addiction have enriched the world of psychiatry. The addicts go through different stages which are inter-linked.

The Over Use of Internet

One of the symptoms of addiction is the over use of the computer. But no one has determined the safe limit for computer use. The computer students and IT professionals use the device for long hours. But for others the limit can be kept at two hours per day. The addicts will not abide by the rule and will declare that computers have become part of their lives.

Ask these questions to yourself, find out whether you have become an internet slave.

1. Are you exceeding the pre-planned time on the internet?
2. Are your relatives complaining about the long hours you spend on the internet?
3. Do you have a strong feeling to continue when using the computer?
4. Do you beg others to give you more time to spend on the computer?

5. Have you failed in your attempts to control the time spent on the internet?
6. Are you trying to hide the fact that you are spending long hours watching internet?

If you give a positive answer you have become an addict.

Withdrawal

Withdrawal symptoms are similar in case of liquor addiction, drugs and internet addiction. Anger, anxiety, sadness and boredom are felt during the withdrawal phase. The addicts may also become restless.

Addiction Oriented Life

Use of liquor and drugs start in a small way. Then the level increases and more kick and pleasure is demanded from them. In the same way internet addicts want to have more pleasure and toxicity from new electronic and computer devices.

Chapter 2

Who Becomes Slaves of the Internet?

There is an element of internet addiction in all of us when we make friends through internet. We become part of 'faceless society'. The members of this society feel that their opinion is always right. They spend more time in solitude. Some teenagers may have difficulty to merge into campus life. Then they find pleasure in the internet. The introverts and extroverts are equally attracted to the internet. The timid find solace away from the realities of life. The smart find pleasure in making friends and opening up new vistas of life. Those who are indecisive and those who are aimless have a tendency to become internet addicts. They find happiness in this new mesmeric world. They feel confident when using a new device and exchanging new ideas.

Interest in the internet becomes harmful when the use of it becomes compulsive. The disease has several medical names. Internet over use, problematic computer use, pathological computer use. Even though the word addiction is removed all these names denote the disease which affects our life related to the computer.

Biological Base

The brain cells communicate with each other using many chemicals. One of the main chemical is dopamine. The Ventral Tegmental Area, Amigdala and Nucleus Accumbens area are situated at the middle of the brain. This is known as the area of pleasure. Dopamine is produced in the area and the person experiences pleasure. Then the person begins to use more and more toxic substance and ultimately becomes an addict. Research says that the same effect is felt in internet addictors. Continuous use of mobile phones also has the same effect.

The study about Chinese students has found that those who make use of the computer for 10 hours a day, 6 days a week have less cells in the brain than the normal students. The brain areas are dorsolateral pre-frontal cortex, rostral anterior cingulate cortex and supplementary motor area. Some parts of supplementary cortex area were affected by the addiction. May be the changes occurred because the brain was trying to adjust to the new situation and to the computer. But these changes reduce memory power, the power to make decisions and persuade a person to move away from the real world to the mysterious world of internet. Patric Wallace, the senior director of John Hopkins University Centre for Talented Youth also supports this discovery. She says that 43% of students who are dropouts from various universities in New York are internet addicts.

The Warning Signals

1. Immersing in the thoughts about internet at all times
2. Internet use to seek more pleasure
3. The failed attempts to control or stop addiction
4. Feeling depressed or restless when trying to control or stop internet usage
5. Breaking the pre-planned time to use the internet

6. Trying to hide from family and friends the time spent on the internet
7. Trying to escape from guilt, depression and anxiety by escaping into the internet
8. The loss of personal relationships, job and education due to the addiction

Those who have the tendency for addiction

Whom do you visualize when you read about an internet addict? It may be a person who sits before the computer for hours together, who is an expert in computer, a young man whom you know well. May be others also think of the same person with regard to internet addiction.

Population Studies

The truth that more than half of the internet addicts are youth is surprising. The youth below 30 are more prone to internet slavery. We should realize the fact that others beyond thirty have problems too.

Many women are prone to addiction. The studies show that men use the internet to gain knowledge, power and position. They also turn to the internet to enjoy pornography. The women immerse themselves in the internet for online friendships. They often spend time on the face book, twitter or social media network and chat rooms and ultimately become addicts. Those who come from broken families or have lost loved ones are more to become slaves of the internet. This tendency is dangerous because it leads to more broken up relationships and grief.

Internet addiction among the middle aged

Internet slaves can be categorized according to age

1. Those below 19   -35%
2. Those between 19 and 30   - 30%
3. Those above 30   -25%

Middle aged people are more prone to addiction than the aged people. The treatment cannot be the same for both sections. Middle aged men immerse themselves in online games. But most women go for the face book or twitter as a past time. The old people go to the internet as a remedy for loneliness.

The All India Statistical Board and the National Information Society states that the teenagers are prone to addiction in a dangerous way. 6.8% youth between 20-49 and 7.9% of children between 5-9 and 10.4% of teenagers between 10-19 have a tendency for addiction. The 20 years olds have a 9.2% chance while 40 years olds have a 4.7% chance to become slaves of the internet.

The malady occurs more in children because their brains are not fully developed. When normal people spend about 1.9 hours on the internet, the addicts spend about 2.7 hours online playing games. The teenagers who use smart phones and become addicts are 11.4% and the 20 years old youth are 10.4% in the case of addiction.

CHAPTER 3

Internet Addiction – The Effect on the Body

Those who sit before the computer for hours together are prone to many health problems. Information technology people suffer from many health problems. They are job-related. The same problems are faced by internet addicts.

1. Carpal Tunnel Syndrome

Median nerve helps us to utilize the thumb and three figures adjacent. They are safe-guarded in a tunnel named carpel tunnel. The tunnel is made of cells and bones starting from the wrist. Sometimes the nerve gets damaged and ligaments become swollen by misuse or over use. This puts pressure on the nerve against the bone and the ability to move finger become less. This produces numbness and pain in the fingers.

The hand should have complete rest when this malady occurs. The wrist may be tightly bound to a scale and prevented from bending. A pain killer or external balm may be used. In case of serious disorders an operation may be suggested.

2. Strain Injury

The strain on the muscles due to constant sitting and bending before the computer may cause pain in the hands and tendons. Some put a lot of pressure while using the mouse and key board. Workaholics are prone to have strain injury because they do not give rest to their muscles.

Body strain, restlessness, numbness in wrist, body adjustment problems, back pain, neck pain and loss of sensitivity are symptoms of this disease.

3. Computer Vision Syndrome

Studies have found that various vision problems arise due to the consistent use of computers. Severe headache, tired eyes, swelling, vision complaints, burning feeling in the eyes and double vision are main problems which effect the eyes. Those who use the computer for more than two hours a day are prone to eye disorders.

4. Radiation

The electro-magnetic radiation is weak and cause little trouble to the body. But keeping in mind the fact that computers have become a part of life recently, long term effects cannot be calculated at present.

Treatment

The effects of computer on the body can be treated to an extent. Instructions are to sit straight before the computer. Turning and twisting should be avoided. The angle of the elbow should be adjusted in a correct way without putting strain on hand muscles. The typing on the key board should be soft and easy. The pressure on the fingers should be kept at a minimum.

Try to relax the body and mind when using the computer. Breathing should be controlled. After each hour the body should be given rest. Take a break every two hours. You can install break reminder software in the computer to remind you time for a break.

Instead of using the computer for long hours resort to writing on paper in the usual way. Do not attend phone calls by keeping the cell phone between ear and neck twisting the body. Multitasking will surely prove hazardous to body and mind.

Computer room

If your job is computer-oriented a separate room is recommended for your use. The room should be airy and healthy. A chair with light adjusting devices and air cushions is a must. The chair should give your back enough support and protect your hips. A foot rest may prove helpful. The monitor should be placed at a convenient height in proportion to the height of the seat. The air conditioner should be kept normal. No unnecessary equipment should be kept nearby. The mobile phone should be switched off.

Key board

The key board should be at a level with the elbow. An easy leisurely angle should be set up for typing. The wrist must be kept straight when typing. The shoulders must be kept relaxed to minimize body strain.

Mouse

When using the mouse the elbow and shoulders must be close to the body. Only the least pressure should be given while clicking the mouse. A pen mouse can be alternatively used. Wireless mouse are also useful to minimize strain.

Monitor

The monitor should be placed just above eye level. A distance of 25 inches is suggested when the monitor is used to copy or typing. The copy should be kept at the side of the monitor on copy stand. This will reduce eye strain and neck pain.

Computer and eye sight

Those who constantly use computers as part of their jobs should get their eyes tested every year. The focal distance from the monitor should be adjusted to the strength of spectacles. Those who use contact lenses should open and shut eyes often. The computer print should be large and legible. There are special spectacles which reduce strong light and glare. The contact reflector screen should be cleaned often. Dark letters in a light background is preferable. There are some exercises which will ease and give strength to eyesight.

Every half an hour eyes should be given rest. Looking at an object 20 feet away for some time will be useful. After rubbing hands for 10 seconds keep the warmth over your eyes. This is a good exercise for the eyes.

Train yourself to breathe in a normal leisurely way. Some breathing exercises are also recommended. These are good for general health and prevent diseases.

Chapter 4

The Online Activities of College Students

Internet addiction among teenagers are on the high. They are often removed by psychiatrists to camps where they live without electronic equipments. The teachers should be aware of the fact that students who use computer constantly may turn into addicts.

- They may make friends on the internet and send email often
- They may download movies and television serials from the internet
- They depend on the face book and social media to gather latest information
- They make use of chat rooms to make love talk

- They are tempted by the prizes and money when they make maximum use of the computer.

Porn Site

Needless search for pornographic information and pictures.

The circumstances by which a college student becomes an addict

1. Students get freedom to use the computer
2. They are in a new world without the supervision of parents
3. They can be free from the inspection of teachers
4. The sites which are harmful are not blocked by the elders
5. Encouragement from college teachers
6. To ease the pressure of studies
7. The fear of the social circle and feeling that they are outcasts.

Some Warning Signs

Most of the parents are ignorant of the internet addiction of their children. The parents should watch for these signs of addiction.

1. Total tiredness
2. Learning disabilities
3. Uninterested in sports and games
4. Keeping away from friends
5. Disobedience and anger

If your son/daughter shows any of these signs there is a chance of turning into addiction. The drug addicts and liquor addicts also have such tendencies in their behaviour. The liquor addicted father, the workaholic mother, the loss of a loved one also lead them to addiction. The counsellor should talk with the addict and find out the circumstances which lead him/her to addiction

How to Deal with Internet Slaves?

If you realize that your son/daughter has become an internet addict, deal the problem in a subtle way using power of reasoning. Some methods are suggested for your use.

1. Discuss the problem with someone you can trust and seek his/her help

2. Give more attention to your ward
3. Place the computer in an open room where everyone can see the user
4. Plan and restrict the time on the internet
5. Persuade your ward to take up other activities
6. Give enough support to them in every way
7. If necessary, seek the help of a psychiatrist
8. It is better to seek help before the addiction becomes severe
9. Educate children about activities that are forbidden

Small children are not familiar with the internet. For them it is a new device. So the elders should instruct them about forbidden activities.

Good Activities

When online if someone asks you to do shameless things immediately come off the site and inform parents. The address of the person should be taken for parents to take necessary action. If they go online without the supervision of elders they should take up the responsibility of guarding themselves. If someone sends you pornographic pictures or talk in an indecent way immediately shut off the computer and inform your parents.

Things you should not do
1. Never give your name and address to a stranger
2. Never receive gifts from online friends
3. Never go to forbidden sites

If your child comes to you with indecent behavior of stranger on the computer praise him for being open and honest. Most children are afraid because they think it is their fault that others are misbehaving. Remove this misconception and give them full support.

There are some people who misbehave on the streets. Some find pleasure in teasing you through phone calls. There are such people in the internet world too. Make your child understand the fact and learn to avoid them.

Self Study of Internet Addiction

A teacher or counsellor should have thorough knowledge of the internet and the tricks children play online. Ask them what they gain and expect from the internet. Notice children who are depressed and anxious and find out the sites they usually visit.

If a young person has to use the computer constantly he/she should know all traps of addiction. Irregular food habits, gambling addiction also come out with the same symptoms.

The serious irregularities you suffer will vanish when you change your habits in using the internet.

How Internet Addiction Affects Life?

1. Sleepiness and extreme tiredness
2. Low performance in exams
3. Moving away from friendships
4. Avoiding social life
5. Feeling restless and uninterested when not on the internet
6. Arguing that internet addiction is not a serious problem
7. Neglecting studies and spending more time on the internet
8. Telling lies about the time spent on internet and going to sites which are forbidden

When a student uses the internet for study purposes he/she should concentrate and venture not to go to other sites. They should participate in club activities, art and sports. They should merge into the social life of the campus and try not to get addicted to the internet.

Persuade the students to use the college library. Convince them that there is more information in books than on the internet. A good library can provide interesting and informative books. Once the student becomes a lover of books, he/she will not even going to the internet.

Those who have overcome addiction should inform others about the malady and try to persuade them to come away from the internet trap. Psychiatrists say that this is more fruitful than other methods. The authorities should hold seminars and symposiums to discuss and inform the student of internet addiction. Every college should have proper counsellors to help the students. The students should have a fixed time for the use of computers. Internet connection is not needed in every computer. There should be posters to warn the students of the bad effects of addiction. Instructors should make them realize that computers are meant only for studies.

CHAPTER 5

Internet and Occupation

Symptoms of internet addiction which affects the job

1. The turn over becomes less
2. The number of mistakes increase
3. The relationship with co-workers will become less
4. Gets irritated in the presence of others
5. Cannot get adjusted to the work place
6. Extreme tiredness
7. Lagging in all types of activities

Seeking Help

Internet addiction is officially recognized as a disease. The company management should take steps to help the addicts.

Ask proper questions

Either the manager or the human resource expert should take up the matter with the employee in person. The employee should be convinced that the company is ready to help him/her to come out

of addiction. The company should ensure them that they know the behavior of all the internet addicts. Convince them that the company has taken up the issue seriously. Do not intimidate the employee and make him understand helping employees is the policy of the company.

Decide whether the employee needs help

If the employee states that he is ready to control the addiction on his own or just denies his addiction, convince him of the bad effects. They should be persuaded to undergo an addiction test. If the test is positive, they may express remorse about the loss the company had to bear because of their addiction. These types of employees need psychiatric help.

Find out the right remedy

The company should appoint experts to provide correct remedies to liquor addicts, drug addicts and internet addicts. Ensure awareness and try to rehabilitate them. The patients should be treated at an emotional level. Their feelings should be taken into consideration.

Control Internet Activity

The company should know how many employees need the computer and internet for their jobs. The time for internet use should be strictly restricted. The computer should be kept in a common place where everyone can see the user. Computers connected with the internet should be exclusively used by the higher authorities. All employees should abide by the rules set up by the company.

The small methods to overcome addiction

1. Restrict time

A plan should be executed and followed for at least three weeks about the use of internet in offices. If more time is spent make an apt plan to remedy addiction.

2. Be patient with yourself

Realize that it will take time to overcome internet addiction. When compared with other problems, the addiction takes less time to be overcome.

3. Congratulate yourself for your efforts

If you are trapped in the internet you will feel shame and anxiety. The escape is not easy. Congratulate yourself for the small steps taken to curb addiction. Saying good-bye to chat rooms and minimizing the time spent on computers is also important.

4. Finding the deep rooted reasons for addiction

You should analyze the deep rooted reasons which lead you to internet addiction. Emotions have to be checked. This will make you take healthy decisions. Life presents us with many choices. What you choose and why you choose are important questions. If you have a clear prospective you will be able to make a wise choice.

5. Receiving the support of loved ones

Your life partner or loved ones will give you support to overcome addiction. They should know that you are trying to restrict time spent on internet. Explain your endeavour and they will give you enough support to overcome addiction.

The symptoms of overcoming addiction

1. You are following the pre-planned time schedule on the internet
2. People are observing a change in your behavior
3. You are conscious of the monthly budget for the use of internet
4. You are regular in your job and inter-related affairs
5. You have redeemed the forgotten pleasures of the past
6. Now you have time to spend with real people other than online friends

7. You realize that you create problems for others including you by your addiction
8. You are not concerned with other sites when you use the internet for official use
9. You have enthusiasm to spend more time with family, take them out or enjoy time with friends
10. When you look back to the days of addiction you feel that you are a changed person now

Chapter 6

How to Realize Internet Addiction?

If internet addiction does not affect your daily life, it may not be a serious problem. But if it changes your attitude to inter-personnel relationships and lead to a break up in home life, the addiction should be taken up seriously. It may also affect your financial situation.

How to find out of the problem is serious?

a) Internet addiction test

Famous doctor Kimberly Young is the first person who systemized a test for internet addiction. It has proven to be very successful and trustworthy at the first stage itself. The addicts are put into three categories: Those who use the internet for less time, who use it in a normal way and those who spend a lot of time.

b) Problematic and risky use screening scale

The famous psychiatrist Gelershick and co-workers assembled a questionnaire in 2012 with 18 items. This scale is used in many hospitals and schools. The research on this scale is continuing.

c) Compulsive internet use scale

Mirkirk of Netherlands Addiction Research Institute with the co-workers developed this scale. It consists of stages starting from zero to fourteen. If the person gets zero on the scale he is not addicted. But if he scores 4 on the scale the problem is serious. The scale shows how many times your attempts have failed when you try to move away from addiction.

d) Mare de-Grefiths test scale

Famous psychiatrist Mare de-Grifith has devised a five item scale to determine internet addiction

1. Importance – How the internet has a strong hold in controlling life and is affecting the thoughts and emotions and behavioural patterns.
2. How the internet has affected the mind set
3. How the addict depends on the internet to deal with his changed mind set
4. Withdrawal symptoms
5. Going back to addiction - after withdrawing from addiction the tendency to go back to old habits.

Internet Addiction Questionnaire

Answer truthfully to the questions given below. If most of your answers are often/regularly you must seek the help of an expert or a psychiatrist.

1. Are you using the internet for more time that you have planned?
a) No
b) Some times
c) Often/regularly
d) Rarely

2. Do you feel close and love the internet than your life partner?
a) No
b) Some times

c) Often/regularly

d) Rarely

3. Do you neglect your day to day affairs to spend more time on the internet

a) No

b) Some times

c) Often/regularly

d) Rarely

4. Does your internet watching affect your job?

a) No

b) Some times

c) Often/regularly

d) Rarely

5. When asked what you do on the internet do you try to defend yourself and hide things from others?

a) No

b) Some times

c) Often/regularly

d) Rarely

6. Have you noted that the output in your job field is affected by your internet habits?

a) No

b) Some times

c) Often/regularly

d) Rarely

7. Are others complaining about your addiction?

   a) No

   b) Some times

   c) Often/regularly

   d) Rarely

8. Do you check your email even before attending to an urgent job?

   a) No

   b) Some times

   c) Often/regularly

   d) Rarely

9. If someone disturbs you while you are watching internet do you feel rage flaring up?

   a) Never

   b) Some times

   c) Often/regularly

   d) Rarely

10. Are you worried of going back to the stage of slavery to internet

   a) No

   b) Some times

   c) Often/regularly

   d) Rarely

11. Are you trying to overcome sadness by spending more time on the internet?

   a) No

   b) Some times

c) Often/regularly

   d) Rarely

12. **Are you of the opinion that life without the internet would be boring and aimless**

    a) No

    b) Some times

    c) Often/regularly

    d) Rarely

13. **If someone needs help when you are watching internet do you ask him to wait?**

    a) No

    b) Some times

    c) Often/regularly

    d) Rarely

14. **Are you spending time till midnight on the internet?**

    a) No

    b) Some times

    c) Often/regularly

    d) Rarely

15. **Are you always thinking of the internet world even when you are not using the computer?**

    a) No

    b) Some times

    c) Often/regularly

    d) Rarely

16. **Are you trying to hide the fact that you are spending most of your time on the internet?**

    a) No

    b) Some times

    c) Often/regularly

    d) Rarely

17. **Do you spend more time on the internet than going out with your friends?**

    a) No

    b) Some times

    c) Often/regularly

    d) Rarely

18. **Have you failed in your attempts to reduce time spent on the internet/**

    a) No

    b) Some times

    c) Often/regularly

    d) Rarely

19. **Do you feel depressed or restless when not on the internet? Do these feelings vanish when you open the internet?**

    a) No

    b) Some times

    c) Often/regularly

    d) Rarely

20. Are you making new online relationships?

   a) No
   b) Some times
   c) Often/regularly
   d) Rarely

Chapter 7

Cyber Sex Addiction

Cyber sex addiction is a serious problem spreading all over the world. Most of the internet addicts seek sexual satisfaction from online conversations. Cyber porn addiction, internet sex addiction and cyber sex addiction are different names for the same disease. The study conducted in Italy says that cyber sex addiction affects personal life, social life, job and sexual life along with family life. Like addiction to liquor and drugs it has a toxic affect. Reports say the addiction is on the increase all over the world. A survey says that 7 out of 10 who indulge in cyber sex are addicted and try to hide their acts. One out of five men who use the office computer watch sex online. One out of eight women use the internet for the same purpose. Men use the scenes to get sexual satisfaction. They are interested in pornography. Ladies indulge in sexual talk.

Cyber sex obsessed men are likely to masturbate while women indulge less in such activity. Pornographic words, letters and stories are often exchanged. Inviting sex partners

through email, sending nude pictures using web cameras, trying to connect with the other person on the phone and sex talk are part of cyber sex. Ultimately the relationship ends in meeting the stranger in person and indulging in sex.

Studies say that women who are over interested in sex ultimately become cyber sex slaves. Some cyber sex prone people are cruel and do unlawful activities. Some are sadists while others are masochists who find ways to hurt themselves. Some addicts find pleasure in watching teenagers and children in pornographic pictures. They also find pleasure having sex with children.

Internet provides a way to connect with strangers. This result in cyber sex activities. It also gives a convenient way to hide your personality, sex, age and religion and job status.

Cyber sex addicts make it their business to widen their web. They find ways to enter into the home page of strangers and change them into pornography. They go into the sites of famous people and prestigious institutions filling them with pornography thus misleading the internet users. Some attract the attention of children by posing as pop-artists. They invade the sites of producers of toys and lead children to pornography.

As most of these sex addicts hide their real personality, it is difficult to assess the extent of this malady. Some studies say that 3-6% of American are sex addicts. They are different from the cyber sex addicts. They do not need the internet to pursue sexual activities. The addicts of liquor, drugs and those who are mentally ill may turn to cyber sex. Cyber-sex crimes were always related to men, but recent studies say that women also can become cyber sex criminals.

The famous psychiatrist Goodman has conducted a study about the behaviour of cyber-sex addicts

1. They repeatedly fail in attempts to get out of their sexual habits
2. They feel guilt and mental pressure when indulging in immoral sex
3. When indulging in pornography they get a kick and the mind and body gets rest and peace

Do you have five of the symptoms among sex shown by the sex addict?

1. Indulging in bad habits even before the malady turns toxic?
2. Spending more time on cyber sex than pre-planned
3. Constant attempts to control or get over the habit or control it
4. Spending more and more time or trying to get out of the habit using more time
5. Continuing this habit which is toxic in doing the job, or while learning or in social life.
6. Turning away from important job assignments social responsibilities and sports.

Reasons

The researchers on cyber sex addiction point out the following reasons for the malady

1. Some trauma which has affected the mind
2. Some change in the neuro chemistry of the brain
3. Some disorder of neural pathways which control the mind set
4. Problems such as depression, anxiety and lack of concentration
5. Trying to assimilate regular sites and adjusting to them

6. Being on outcast from the society in the early stages of life
7. Loneliness, boredom or severe ego consciousness.

A mixture of all these feelings

Those who have inferiority complex, ugliness, sex problems and are homosexual have a tendency to become cyber sex addicts.

The sexual experiment done at a young age develop into an uncontrollable urge to have cyber sex. Some addicted people are college students. Some try to escape from the severe pressure of life by going to porno sites. Some become sex addicts because they cannot bear the problems in life.

Our eating habits, sex habits and addictions effect the brain cells. Breathing and functions of the heart are not controllable. Likewise the sex urge is also uncontrollable. Endorphins, Encephalin and Catecholamine are produced in the brain without conscious effort.

Hypothalamus is the controlling centre of sex urge. The nerve centers, hormones and emotions join together to make the Hypothalamus work. Attraction, companionship and libido are the basic emotions of the brain. It is the sexual urge that made the mankind multiply. Estrogen and androgen are the hormones which lead us to procreate.

The second emotion persuades the person to seek relationship with another then hormones dopamine and Nor-epinephrine are produced and the production of serotonin is decreased. Dr. Fisher, the famous psychiatrist is of the opinion that with the disappearance of serotonin the person is attracted sexually to another. The production of dopamine and epinephrine increases at the first stage and lessens at a later stage. When serotonin hormone reaches the normal stage, companionship develop among the sexual partners. It takes 6 to 18 months for the sexual relationship to reach this stage. This leads the person to take up the partner and family in a responsible way. The feeling of closeness and faith is necessary for a fruitful companionship. Then the two experience peace and happiness. The chemicals Vasopressin and Opioids are produced and provide feeling of contentment. The mothers who are breast feeding feel a strong bond to the baby because of oxytocin produced by the brain. Imagination, sexual urge and rest after a sexual act are part of sharing the feelings of love. The cortex part of the brain leads us to imagination.

All the changes in mind set are due to the working of chemicals produced by the brain. They become toxic in excessive quantity. The person who seeks pleasure needs these chemicals regularly and at an increasing rate. The sex addicts continuously peruse sex and the brain has to produce more and more these chemicals. Liquors and drugs also have the same effect. The activities of these chemicals are the same in all addicted people.

Cyber Sex Addiction, the Ultimate Result

a) Social
The online sex addicts move away from family and friends and become social outcastes.

b) Emotional
Addicts are always fearful of exposure by the society. This creates extreme mental pressure. They feel guilty and shameful because they are breaking trust and morality. They feel bored, tired and hopeless. They are led to depression, suicidal tendencies, inferiority, extreme anxiety, loneliness, crash of opposing thoughts, useless thoughts and remorse. These conditions ultimately lead them to suicide. The cybersex addicts may come from broken homes. The parents may also be cyber-sex addict.

c) The effect on the body
Cyber sex addicts develop abrasions on sexual organs. Cancer of the womb, herpes, warts around the rectum are seen in addicts. If expert help

is not sought on time, these will worsen or even lead to death. Those who watch pornographic videos while driving usually end up on accidents.

d) Laws against cyber sex

Sexual crimes/pornographic calls on the phone, exhibition of genetic organs, peeping to watch the sex activities of others, or their nudeness, rape, prostitution and misusing children for sexual satisfaction are all against law. Cyber sex addicts often indulge in these crimes and are punished. Some even lose their job when exposed.

The effect of cyber sex on jobs and economic affairs

Prostitution: Cyber sex and sex talk over the phone effect your home budget. Many addicts lose all their money contesting divorce cases. They have to give money to the advocates and pay alimony. They may also lose their jobs and become paupers.

Moral effect: Lack of faith in God, the feeling of remorse and guilt will lead to mental breakdown.

How it affects children?

a) The children of cyber sex addicts also happen to see dirty scenes on the internet
b) The children are left alone with no attention from parents
c) The situation of broken home affects the mindset of the children

When parents are angry the children may take part and their mental peace is affected.

The effect on the life partner and family

Spending unending hours before the computer has many bad effects. Dr. Cooper, the famous psychiatrist says that cyber sex addicts spend more than 11 hours a day on the internet. Thus life partners feel left-out, neglected and become victims of anger. The family also feels neglected.

The partner suffers from inferiority complex when he/she discovers the online sexual relationship. They often react in a violent way. Some with vengeance may turn to other relationships. Some may commit suicide because of depression, hopelessness and inferiority complex. Faith is lost in each other. This is a dangerous situation in marriage. Many unruly love relations are developed. The addicts try to bring others also into the unruly state. Cyber sex addicts often show rage to their partner. They often imply that they have turned addicts because the partner is at fault.

How cyber sex affects sex life?

The cyber sex addiction affects the partner in a serious way. 63 out of 94 partners are of the opinion that they have unsatisfactory sex lives. A study says that 52% of cyber sex addicts are not interested in sexual activity in real life.

The continuous problems

1. The cyber sex addicts avoid sex in real life by finding excuses like tiredness, or that children might hear their sexual activity.
2. The partner feels that she/he is good for nothing and cannot compete with the person on the internet. They suffer from extreme rage
3. Cyber sex addicts think only of their own pleasure. They fail to understand the feelings of their partner and to give them pleasure.
4. They blame their partner for the failure of sex life.
5. They insist that the partner should play along with the whims to give pleasure
6. The partners may feel that they may contact sexual diseases from the cyber addict.

Is cyber sex an extra marital affair?

When a partner indulges in cyber sex many problems arise in the family. There are no definite norms for restricting cyber sex. A study says that 59% of college students feel that oral sex is not unlawful, 19% feel anal sex is not an unlawful activity.

Treatment

With the development of technology, treatments for cyber sex activities have become complicated. The use of computers has become a part of life. People who have jobs connected to computers and students who have to use computers constantly have a tendency to become cyber slaves.

Therapists first gather information through the partner or family members of the addict. They also try to find if the patient has some other addiction. Counselling is needed for the whole family and friends of the addict.

There are three types of therapy

1. Therapy for the individual
2. Group therapy
3. Cognitive behavioural therapy

Some serious cases need medication when therapy itself is not effective. The Sex Addiction Anonymous and Love Anonymous and several honorary counseling centers give effective treatment to cyber sex addicts. There are also support groups and sites which help the addicts and their families.

The object of therapists

1. They should convince the partner that competing with the cyber sex partner is impossible
2. Persuade them to take right decision

3. Convince the partner to stay away at a safe distance from the addict
4. Inform them of the institutions which will offer them help.

It is best when both the partners join together to solve the problem. There are soft wares which block the use of pornographic sites. Alwin Cooper, who is an expert in cyber sex studies suggests this method as the best.

How to decide the extent of addiction?

There are many methods to determine the scale of addiction. We must analyze them.

1. Sexual compulsive scale

There are 10 to 40 points on the scale. The therapist asks questions to the patient and determines the points on the scale. Those who score 18 points are not addicted. The points between 18 and 23 show a little tendency for addiction. Those who score point between 24 and 29 have become addicted. Those who score more than thirty points are dangerously addicted to cyber sex. This method is very reliable.

2. Cyber sex self test

This test developed by Dr. Kimberly enables the patient to answer 10 questions and self determine his addiction on a scale.

3. Sex help

Internet sex screening test with the help of the website com.

4. Society for the advancement of sexual health

This screening test was developed by Rob Vis after much research. This helps to determine the nature of addiction.

Cyber sex addiction, internet addiction, mobile addiction and video game addiction have come up with the recent advancement of technology. These show how human beings become slaves of machine and technology.

The advancement of science and technology has both positive and negative effects. Sometimes we may not be able to decide what is normal and what is addiction. Technology has raised our standards of living. It has made enormous change in the field of health and longevity. But the negative effects have made changes in body and mind. The knowledge of various addictions makes us aware of the malady and to beware of situations which lead to addiction.

Chapter 8

The Cyber Threat – Problem of the Future Generation

Like the invention of atom bomb and hydrogen bomb technological inventions have both positive and negative effect. Not only elders but also children fall prey to the negative effects. Email, chat rooms, mobile phone, mobile phone camera and websites are all used to intimidate children. Even the youth try to catch preys in their webs. This tendency has spread in America, Japan, Canada, Britain and India. The society has not taken up this problem seriously.

Threatening was considered child play until now. But for the past decades the problem has become severe. The threats which arise in school are taken up in the hostels. The victim is attacked because of a strong inner urge. The violence arises without any reason. School children get into fights boxing or kicking, spitting or calling bad names or using bad words when two children fight. The onlookers do not interfere. They are afraid that they may become next victims.

The new method of fighting is with cyber means. Children use email, chat rooms, mobile phones and mobile cameras and websites just like the elders to intimidate others. Degrading messages on the mobile phone, emails which are threatening, and private secret messages are used to intimidate children. They create a bad image of a person and post them for the opinion of others. The child who is a victim may respond in the same way. Quite recently a mobile picture of a girl changing her dress in the bath room was spread widely through internet.

When Cyber Threats Become An Epidemic

The habit of threatening is wide spread among school children. Many studies say that at least one of six school children are victimized. 40% adults remember victimization in school days. The study conducted in Britain Children's National Home found out that one out of four children were threatened via mobile or email. The principals and teachers say that this behaviour is wide spread. The mobile habit spreads from adults to children. Both boys and girls suffer from intimidation when boys are prone to fight face to face girls are threatened through cyber means. Girls tend to use mobiles constantly and are easy to be victimized. Primary students are threatened in person while high school students are confronted via cyber means.

The victims become depressed and afraid. They feel acute anxiety and fall into inferiority complex. They feel that they are good for nothing. They become unable to express themselves and their feelings.

The cyber way of threatening is more vicious than a direct one. More people come to know the threat. The email messages and websites are often open and wide spread. It is difficult to find the pursuer as the messages are anonymous. This can happen to anyone and everyone and is very dangerous.

Prevention

Researchers have studied this problem for two decades and suggest four methods to curb this tendency.

1. Revealing the extent of this malady

The school management should willingly admit the problem of cyber threats. The teachers, parents and children must be aware of this. Classes should be conducted to teachers to reveal the methods of intimidation on the internet. The parents should be aware of the methods by which children send emails to threaten others. Teachers and parents should have an open talk about cyber ways of intimidation.

2. Make a policy

Each school should have a definite policy to prevent and punish cyber threats. The policy of one school might differ from another school because of certain circumstances. If the policy proves to be ineffective a new policy should be adopted. The change in policy should be informed to students and parents. Some schools adopt the same policy which is followed by the society.

3. Supervision

The supervision of elders is most necessary to avoid cyber intimidation. When under supervision the fights on the play ground become less. In the same way cyber threats become less under supervision. Teachers should make sure that the students are not going to pornographic sites. Parents should be watchful when children use the computer. Some parents are unable to do this because they are computer illiterate. The children are smarter than them. The school authorities should educate the parents of technological devices misused by children. When the children come to know that parents are competent, they are less likely to be mischievous. They will also seek the help of parents when they are in trouble.

4. Prevention of cybers threat

There are two main ways to prevent cyber threats:

a) Social plans
b) Study materials.

The tendency of the majority of children who look on silently when two children have fight is dangerous. Children should develop empathy.

Those who are empathetic will respond and also try to prevent fights. The same is applicable to cyber threats. Each school should have a help desk and appoint counsellors to help students. The rules of morality and empathy should be taught through moral stories.

The Role of School in Preventing Cyber Threats

Most of the teachers and parents are of the opinion that severe punishment should be given to those who repeatedly indulge in cyber threats. Most of the children are victims but only very few approach parents or teachers for help. They think that parents will not consider it a serious problem or will blame them or make the problem worse. They fear that computer use and mobiles may be forbidden. Teachers and parents should have trust for their ward and take up the problem seriously. Measures must be taken to maintain secrecy. Correction methods should be taken only after joint consultation with parents and children.

The problems of school children are limited to school premises. But cyber threats can follow you without any regard to place and time. Intense fear is felt because the threats are anonymous. They reach more people through cyber ways. This type of intimidation is a very serious problem which is on the increase in the modern world.

Chapter 9

Cyber Crime

The crimes done by means of computer, internet, cyber space, world wide web (www) are all considered as cyber crimes. Email fishing is a crime when your bank account number is asked and money is drained. The credit card information is also taken for this purpose. The criminals crash into the internet to get unlawful information. They may do spy work in industries. Crime against children are crime that make use of children, kidnapping, sexual harm to children, spreading viruses. The new generation indulge in more unlawful and wasteful search of the internet. Cyber crimes are increasing day by day all over the world.

Crimes can be definite or indefinite. Some can be found out easily while others are difficult to find out. Internet crimes are classified into two.

1. Crimes done while making use of the internet
2. Crimes which are conducted only because there is access to the internet.

Email fishing, making unlawful use of brand names of companies and institutions, infiltrating viruses or web sites are crimes of the modern age. The people are cheated when they give their bank account numbers or pin numbers of ATM cards when criminals ask for them for verification. They are convinced that the bank is contacting them for the information. Many people have fallen prey to this type of cheating and lost their savings.

Cyber Laws

The Indian Penal Code has not yet defined cyber crimes. Cyber laws are made for crimes connected with the internet. Intellectual property, privacy, publication rights and limits of power are defined by laws.

Even after amending cyber laws in 2008 the Indian Penal Code does not mention cyber crimes. The Right of Information, the guarding of information and right of information related to computer, unlawful infiltration into information, changing and destroying information are mentioned as criminal activity in IT law section 2(b).

All the cyber laws are for controlling criminal activities by using computers.

Statistics

Internet crimes are increasing at a high rate all over the world. In America about 10000 crores of dollars were lost through internet fishing. As per the data maintained, since its inception 3,17,439 cybercrime incidents and 5,771 FIRs have been registered up to February 28, 2021 in the country. Secret information was gained by infiltrating into the accounts of 500 companies and 80% of these had financial loss. The American investigator agency, FBI conducted a survey of cyber crimes. It assessed a total loss of 4000 crores through criminal use of credit cards. About 20% of these crimes were done through the internet. The statistics show the urgency to prohibit cyber criminal activity.

Hacking

There is no mention about hacking in Indian IT Act. Hacking is the infiltration into information or source of information in a criminal way. An expert computer programmer who uses his skills for criminal activity is known as a hacker.

Law and Punishment

The amendment of 2008 of IT Act 43(a)66 section and laws of Indian Penal Code (1860) 379 and 406 are meant to punish hacking. If proved guilty the hacker may get three years of imprisonment along with a fine. The fine may come up to 5 Lakhs. The crime is bailable. The case can be withdrawn through the intervention of the court after discussion with both parties.

Data Theft

Desk top, computer, digital information devices like flash drive, iPod, digital camera are the main devices of office workers who drain essential data from the computers. Wikipedia is of the opinion that this type of crime is on the increase on email, web pages, USB and DVD storage which can be tackled without destroying information. Downloading data without the permission of the authorities and computer network personnel is a crime according to Indian IT Act section 43(b).

Law and Punishment

Section 43(b) and section 66 of Indian IT Act and section 370, 405 and 420 of Indian Penal Code are meant to prevent data theft. It is a bailable offence. The crime is given a chance for mediation by the court.

Spreading viruses

Computer viruses can be inserted to make little changes or wide spread changes according to the plan of the hacker. The secret information in one computer can be sent to a third person and destroyed by the use of viruses. After a virus attack the whole operating system has to be changed. The viruses of past years were not able to cause much damage. But now just at the opening of a site the destroying files installed appear and spoil the system. The viruses then spread to other computers.

Laws and Punishment

43c, 66 and 43e sections of Indian IT Act and IPC section 268 are meant to punish the crime of spreading viruses. The crime is bailable and can be discussed through the intervention of the court.

Indemnity Theft

Wikipedia defines identity theft as misrepresenting a person for financial gain. Copying another person's electronic signature, password and other identifying phenomena and using them to cheat a person are crimes. The section 66c of Indian IT Act is applicable in the crime. The person who is victim of identity theft becomes responsible for all the misdeeds of the thief. Although these criminals do not resort to murder, identity theft has intimidating effects. The effects are beyond imagination.

Law and punishment

Section 66c of Indian IT Act and IPC section 419 are meant to punish the criminals who indulge in identity thefts. They are liable to get bail and the case can be discussed by mediation of the court.

Email Spoofing

Changing the identity of the person who sends email is a crime which is known as email spoofing. It can be done by tampering the password in the computer. The hackers hide the actual address and get access to money and information by internet fishing.

Solution to these problems

There are laws to punish cyber crimes but the problem is that they cannot be practiced effectively. The Indian IT Act and Indian cyber laws are exercised to prevent cyber crimes. The American Secret Agency FBI has prepared online crime registers to agencies which are on the lookout for job personal. Internet criminals are rarely brought before the law.

Hackers probe into computer from nation to nation. They change their email addresses constantly when indulging in spreading viruses and email fishing. Every person should stay alert about these unlawful practices. They should be careful when they open emails from unknown sources. The password must be guarded with utmost care. The use of antivirus and antispyware software should be made a habit. Anonymous email messages and programs should not be opened.

Digital India have become a soft target for criminals as country recorded a huge increase of 63.5% in cybercrimes cases in the year 2019, shows the National Crime Record bureau (NCRB) data.

The NCRB's data stated that 44,546 of cybercrimes were registered in 2019 as compared 28,248 in 2028. The data showed in 60.4% of cases, registered fraud was motive followed by sexual exploitation (5.1%), and causing disrepute (4.2%). Highest number of crime cases were registered in Karnataka (10,020) followed by Uttar Pradesh (11,416), Maharashtra (4,967), Telengana (2,691) and Assam (2,231). Delhi alone accounted for 78% of cybercrimes.

As per the data, in metropolitan cities a total of 18,372 cases were registered, showing an increase of 81.9%. The data also stated that maximum cases (13,814) were registered under computer related offences (Section 66 of IT Act). More the legislation, more frauds will be highlighted, that's why numbers have spiked up so much.

Chapter 10

Prevention and Treatment

Most of the people can control addiction by their own effort. Content control software, cognitive behavioual therapy and counseling are helpful in minimizing addiction. Inter time is limitless and irresponsible people become full time addicts.

The addicts are persuaded to spend more time for sports and games. A group of psychologists and psychiatrists join their efforts to curb addiction.

How to get over internet addiction?

1. First you should admit to yourself that you have become a slave to the internet
2. Realize that thousands of people like you are affected by the same malady.
3. Try to indulge in games not connected to the computer. Sports, club activities, music, dance and social work make you shift away from the computer.

4. Take more interest in your studies and gather more information from books. The rule is applicable to teachers also. Even if there

is no exam on the next day, learn day to day lessons in an orderly way.
5. Avoid chatting and help your parents in cooking and other household activities. This will make you feel good and give you confidence that you are not slave of the computer.
6. Visit places of natural beauty where internet connection is not available. Visit shopping malls or ice cream parlors with friends.
7. Family members should gather around the dining table and share food. A bad habit has developed where each family member has food according to his/her convenience watching TV or mobiles.
8. Self regulate the time spent on the computer. The computer and laptop should be moved to an open place so that the temptation to spend more time can be controlled.
9. Instead of sending messages to relatives and friends contact them over the phone. Spend at least three hours with friends. Try to do homework with your classmates.
10. Set up the alarm to remind you of time before opening the computer. Do not ignore it when it sounds. You can also set up a stopping device in the computer itself when time is up.
11. Avoid the habit of taking food while sitting before the computer.

Some small tricks

1. Make use of the computer in the college library. Thus you can avoid pornographic sites and limit time. The library provides good informative books. You can limit your time on the home computer.
2. Make a list of the pleasures you get when you limit time on the computer.
3. Visit a park or beach and become close to nature.
4. Try to lessen 5 minutes from your time on the internet every day when you use the computer. Set an alarm at a specific time to stop. Make a definite plan for the use of internet.

5. If you need sites like Wikipedia constantly, download the site and save it. You need not go to the site often.
6. There are many addicts who are irregular in their sleeping habits. This is unhealthy. It will also affect you personal life.

Some warnings

1. Try to limit the time spent on computer for school projects and college activities.
2. Give rest to your eyes and muscles after every 15 minutes spent on the computer. Making use of the mouse and keyboard for a long time may cause pain in the hand muscles and end in carpel tunnel syndrome.

The Next Step

If the addiction cannot be controlled on your own seek the help of a psychiatrist or a psychologist. Depression and anxiety problems can be solved through medication.

Calculate the time spent online

Note down every day the time you are spending on the internet. This written account will show you whether you are turning into an addict.

Making use of time fruitfully

1. Instead of spending time on the internet make it a habit to go for sports and games.
2. Record the time you spend on the computer in which days, time and places move away from them.
3. Make a list of the jobs to be completed, the places where you have to go and set a time for internet usage. Set the alarm to remind you of the time.

4. Calculate the time you can spend on the internet for a week. By this restriction you can cut half the time you spend on the internet. Do not spend time uselessly for searching information.
5. Move into social life and mix with people and seek their help.
6. Tighten family ties and make friends. Do some fruitful work for the society. Try some other activity which will give you pleasure.
7. To move away from online friends seek the help of clubs and social welfare groups.

Discovering that you have become a slave

When you are about to open the computer ask yourself to complete these sentences.

1. I feel that I am .... when not going to the world of internet. (bored, listless, degraded, depressed, anxious, angry or pressurised).
2. Next fill up the sentence

When I go to my dear internet world

I feel I am ....

(Peaceful, energetic, happy, hopeful, respected and loved by all. I feel I am a super power in sexuality).

If you analyse the answers you will find out the reasons for spending more time on the internet. The answers are important in the treatment of addiction.

Use positive reminder cards

1. Make a list of your main problems connected with the internet
2. Make a list of the gain you will have if you move away from the internet
3. Keep both the lists in your purse

4. Read both lists before you open the internet and make a decision
5. When the rate of addiction reduces, your problems with job, friends and relatives will vanish. The sleep disorders will diminish. You will not become a social outcaste

Make the right move

A person with clear prospects in life can overcome any addiction. Try not to sit before the computer until midnight. Maintain a fixed time like 10 or 11 pm for sleep. A person who has definite plans in life can achieve many heights.

Beware of negativity

The addicts of liquor and drugs will not admit that they have become slaves or that there is something wrong with them. In the same way internet addicts will not admit their state of slavery. You can find out their attitude through listening to their negative words. 'Please don't disturb me', 'I am doing necessary work using the computer', 'This is not an extra-marital affair or sex play', 'I am not immature in my behaviour', 'After all these are only some words on the computer'. 'when I go to the chat room all my problems vanish' are some of the excuses that addicts use. You may remember hearing these sentences from your friends and relatives.

The problems the addict faces come out as excuses to find refuge in the internet web. The famous psychiatrist Dr Young has made a list of problems which make a person an internet addict.

1. Loneliness
2. Lack of adjustment in marital life
3. Work pressure
4. Boredom
5. Depression

6. Money problems
7. Disability or difference in body structure
8. Anxiety
9. Trying to overcome other addictions
10. Lack of participation in social life

Dr. Young points out three stages on the way to addiction

1. First stage – regularity

The first stage when you know about the computers thoroughly and are familiar with all the sites. You begin to use them regularly. You come to like one or two sites. You become a familiar figure in the chat room or news group. You are beginning to enjoy this new status and recognition.

2. Second stage – seeking support

You rely on the internet to solve the problems in real life. You find one or more friends on the internet. You feel that it is very reliable. It is more supportive and attentive than real friends. You begin to move away from people and neglect your duties in life.

3. Third stage – escaping

You feel peace and happiness when you spend hours on the internet. Then you are devoid of the pressure of life and loneliness. You escape tragic thoughts and negative feelings. But you must realize that there are real friends and support groups in life. Cultural activities may bring you back to the real world. It may change the feeling that you are an outcaste from the society.

When you begin to analyze the negative approach, the problems in real life, and how you are trying to escape then you find that you have reached the third stage of addiction.

Get over your loneliness

Make use of the positive gains that you have achieved through the internet. Make friends in all walks of life without regard to caste and creed. The sharp attentiveness and intellectual gains achieved through the use of internet can be made use in real life in a productive way.

If circumstances are the cause of loneliness try to overcome the situation by joining clubs or cultural groups. Make a thorough change in your job and other activities.

The accident victims who have to stay bed ridden for a long period have a tendency to become internet addicts. Changes in life regarding occupations, new relationships and educational methods may turn you into an internet addict.

You should share your strong feelings and thoughts with loving faithful people and try to overcome them. At first you may find it difficult to place your worries.

At the next stage you may be able to self treat yourself.

Seven signs of internet addiction

If you have doubts of your partner's addiction to the internet, ask yourself whether you have noted 7 signs in him/her.

1. Irregularity in sleep
2. Personal changes
3. Disinterest in sex
4. Yearning for privacy
5. Neglecting day to day affairs

6. Habit of telling lies
7. Drifting away from marital bonds.

Go through these seven steps of communication to find out the mind set of the addicted person.

1. Decide aims

Decide whether you want your partner's addiction to be controlled or put an end to. Come to an agreement whether both of you need psychiatric treatment.

2. Find time

The addicts are irritated when they are busy with the internet. They are reluctant to speak. Find a time to talk in person

3. The important point

Unemotionally explain the pain you feel by his/her addiction and how it affects you.

4. Try not to blame

Try not to blame your partner by using sentences like these 'you are immersed in the internet for a long time at night and I feel neglected', 'my offers of love are rejected'. The practical approach like 'you are not giving me enough attention because you are on the internet at all times' may prove effective.

5. Listen with sympathy

Listen to your partner with attention and respect. Be lenient in your approach for some time. Have an open mind to listen. Have sympathy to his/her problems.

6. Be ready to find negativity

Be firm in your decision that the partner should undergo change. With further discussion try to overcome his/her negativity.

7. Other ways

If a direct approach fails, a detailed letter or email may be sent to the partner about the addiction and your feelings. Do not get disheartened. The help of a psychiatrist is most necessary.

Chapter 11

Mobile Phone Addiction

Mobile Phone, Use and Misuse

One of the most important inventions which changed the life pattern of human beings is the mobile phone. Half of the population of the world uses mobile phones.

India's digital journey is one of exuberances. The country had the world's second-largest internet population at over 483 million users in 2018. Of these, 390 million users accessed the internet via their mobile phones. Estimates suggest that this figure would reach over 500 million by 2023. The same person may use two or three mobiles. This has become a new trend.

People prefer to use mobiles instead of land phones. The new generation have a craze for mobiles. The use of mobiles is wide spread even in developing nations where there are no strong government structure.

With the invention of mobiles the use of personal computers and laptops for internet connection has become unnecessary. Mobiles are also used to take photos, videos and email communication.

Internet and Mobile Abuse

A new culture has developed with the wide spread use of mobiles. The social connection is maintained through mobiles. Sending WhatSapp is a favorite way of communication. People prefer to have smart phones which provide all conveniences. The market value of these services become high.

Mobiles with new apps have become the fashion of the day. College students use them as show pieces and compete with each other owning latest models. They sometimes indulge in criminal activities or unlawful practices to get money to buy a new mobile. This painful truth comes out when you analyze the deeds of young people.

Mobiles have positive and negative effects. It creates disturbance in the class room or get into the privacy of others. The ring tones kept loud are disturbance during funerals, movie shows, play grounds and occasion of marriages. Mobiles often lead people to pornography. Loud conversation over the mobile is a disturbance in book stores, libraries and holy places. Many airlines have prohibited the use of mobiles when airborne. The radio waves and messages may be affected by the sound waves of mobiles.

Many have discarded their old small phones and bought smart phones and connect them to the internet. India will have 966 million mobile users by 2023. Two in every three users are expected to have a mobile phone by 2023, while one in two users will have a small phone. Many mobile apps have become part of life. Google player has 2.8 million users while Apple application store has 2.2 million regular users. Use of Gmail, Google map, You Tube and Google and android applications are most popular. They all have world wide application.

Social network is one of the most popular sites of mobile phone users. Statistics say that 1.3 million users send messages by What Sapp every month. Face book users come next. Other message applications are Q-mobile, V chat and Skype.

50% mobile users indulge in watching movies and videos online. Apple stores and ten cent seem to be most popular. Kivi and Netflix come next in popularity. Music streaming apps are used to enjoy music. In recent years online shopping has spread far and wide. Almost all things in the market are available for online purchasing. A survey shows people have turned to smart phones to purchase things online. They save money and get gifts. Railway tickets, reservation etc. can be done through mobiles. Purchase of home appliances and garments are made easy. One study says that 70.9% like to do shopping online. An average of 4000 rupees is spent every year for online purchases.

Mobiles are intended for our use and benefit. It should not be used to disturb the peace of others. The wise use of mobiles will help you in day to day life.

The history of mobile phones

1876	Alexander Graham Bell invented the telephone
1947	Ben Laboratory put up the idea of cell communication
1973	On April 3rd, a system manager Dr. Martin Cooper spoke to Dr. John S.Engel on a portable cell phone he had invented. Dr. Martin cooper is regarded as the father of cell phone.
1977	Ettikettle and Benlab together put a prototype of cellular system
1979	NTT of Tokyto, Japan began the production of cellular system.
1982	Nokia produced the mobile phone named 'Nokia Mobira' Centre which weighed 21 pounds and was intended for the use in cars.
1984	Cell Lab invented 'Call Hand Off' system where one could connect with a number of cells by using the mobile system.
1990	2G technology was presented. The first 2g call was in America
1991	The first GSM network 'Radio Ginga' started in Finland
1993	Text message began to be used (SMS)
1995	Text messages began to be used in China and Japan for commercial purposes
1998	Blue tooth technology was invented
2000	First smart phone 'Kysare' in the market
2008	Steve Jones presented I phone 3G.

Smart phones have replaced watches, alarm clocks, letters, land phones, cameras and torches.

Pew Research Centre discovered that sending messages through SMS, What Sapp and V chat are increasing day by day. 75% of the world population make use of SMS to exchange news, 72% of Japanese use mobiles to take photos and videos, while 58% of Egyptians show the same tendency. In Israel 47% make use of internet use through mobiles, 46% of Japanese and 43% of Americans are of the same habit. The app that has maximum use is the social network. Even the ten year old are competent in their use of social network.

Misuses of the mobile cannot overweigh its good uses. In an emergency mobile phone is your best friend. SMS is more easy and reliable than letters which reach only after some days. Games, music and news flow from your finger tips. It gives you pleasurable activity. It strongly binds the social network. Purchase of movie tickets.

Chapter 12

Misuse of Mobile Phones

Misuse of mobile phones have spread all over world like an epidemic. It causes mental pressure, anxiety and addiction. The teenagers suffer from mental pressure, anxiety, sleep disorders and tiredness due to mobile addiction. The new generation thinks that smart phone is a part of fashion. They use them to enjoy music, to get access to pornographic sites and to send shameless messages and disturb others.

When you have lost concentration and peace of mind because of the misuse of mobiles you have become a slave. Then you show these or more signs.

1. Lack of concentration to do anything else
2. Little or severe restlessness
3. Inserting loud pop music or caller tunes on the mobile
4. The feeling that there is a person actually talking to you
5. The fear that the other person would put an end to the conversation
6. The fear that the other person will leave the place
7. The fear that the battery of the phone will be exhausted.

I. In attention blindness

Inattention blindness is the discomfort you feel when using the mobile in public places and while travelling. It affects your attention. The users of land phone do not experience this discomfort.

1. The use of mobiles inserts an intense pressure on body and mind
2. At the bus stand or cross roads more strain and attention is needed
3. When crossing the road your attention is diverted and you invite accidents
4. You cannot respond to the sight of an accident on the road
5. Others are irritated because you talk over the phone in a loud voice and you are not aware of surroundings.

II. Caller hegemony

Caller hegemony is the feeling of connectivity felt between two persons even when they are not on mobiles.

III. Cognitive lock

Using the cell phone in public places affects your attention and pressurizes you. It goes against civic sense. The laws forbid you from the use of mobiles while crossing the road and rail tracks.

E-Waste

The main content of e-waste are the mobiles discarded. As technology bring into the market new phones, the old ones are thrown away. Mobiles contain toxic waste such as lead, zinc and mercury. The use of bromine to prevent the mobiles and computers from fire is very harmful for the human body. The e-waste which is discarded in public places enter into water ways and ultimately affect our body. The challenge that the electronic industry is facing is how to destroy e-waste.

Cyber Crime

Cyber crime using mobile phones are on the increase. Mobile camera, multimedia, blue tooth technology all make criminal activity easier. The word cyber is derived from Greek Kubernets, cybernetics. Cyber space, cyber homes, cyber hate are some of the words related. The electronic language uses cyber to denote control.

Psychological issues

The over use and misuse create two types of psychological problems in the new generation.

1. No-mo-phobia (No mobile phobia)

Some are fearful that the battery of the mobile may fail. Some feel anxious about mobile cards and network failure. Place where net work is not available creates anxiety. One way to overcome the fear is to keep the battery charged and pre-pay the sim card. The second way is to keep away from the mobile and thus prove to yourself that it is not very essential in normal life.

2. Ringxiety (Ringtone anxiety)

Ringxiety is the anxiety to listen to the ring tone at all times. Some feel that the mobile is vibrating. Some check their mobiles for missed calls and messages very often. This is a sign of anxiety.

The militants all over the world make use of mobiles to spread their ideology and to intimidate victims. Some strong groups have their own secret network.

Mobile phone addiction becomes worse every day. They are meant for communication or gathering information. They are not meant as a pleasure giving device. It is not part of the new fashion too. The addicts must realize these facts.

Misuse of mobiles by the youth

Teenagers and even elder people use mobiles, for long hours. The drivers use them while driving. The elders use them during office hours and students use them in class rooms. They feel that it is alright to send text messages even when conversing or attending meetings. Most of the teenagers own a smart phone. They send more than hundred messages every day. It amounts to 3000 words per month. The addiction leads the youth to loss of attention, alertness, memory problems and low performance in exams. The efficiency of the youth decrease in the job fields. Sleeplessness, restlessness, anxiety and depression, attention disorder are the outcome of mobile addition. The mobiles must be kept switched off during study time. There should be strict rules for the mobile use in the offices. Neuro imaging studies declare that the constant use of mobiles have an effect on brain cells which cause a state of intoxication.

Chapter 13

Computer Game Addiction

Many parents have the same complaint. My son/daughter is in the tenth standard and is addicted to mobile games. They often become violent if asked to come away from the game to study. They threaten to stop their studies or run away from the house. These are clear cut signs of addiction. This is the epidemic that has effected the new generation. Some of us allow the child to use mobile for some time.  They find new alternative games to play. Slowly they become addicted to the game. The parents may not realize this for a long time. The World Health Organization has declared this game addiction as one of the new generation disease. The name given is 'gaming disorder'. This is one of the serious negative results produced by the progress of technology.

Slaves of the Mobile

The addicted children prefer to play on the mobile more than anything. They neglect sleep, food, studies and friends. They know the bad effects but are not able to control the yearning.

If the habit continues for 12 months it can be called game addiction. Statistics are not available about game addiction in our country. But in foreign countries at least 85% between the age 8-12 are slaves of computer games. 15% of those who regularly play game turn to be addicts. Adults above 35 years of age play games for monetary benefit. Recently it has affected the ladies also.

Is it a disease?

This game addiction is similar to addiction of toxic substances like liquor and drugs. The intensity of addiction can be calculated on a scale named CAGE. The same scale is used for liquor addiction. C means to cut down the time spent on games. When addiction is put to a stop many problems arise. A stands for annoyance. The victim is intolerant and feels annoyed when asked to stop the game. G stands for guilt. The addict feels guilty but spends more time playing games. They will not listen to themselves. E stands for eye opener. The first thing the addict thinks in the morning is what game to play that day and whether there are any new games introduced.

Two types of games

1. Standard game

This is the usual game played with the use of CD. There is definite aim such as finding the lost princess. The gamer should get maximum score within a limited time.

2. Multi-player games

Type two games require more players. They make steps on their own and go forward. Sometimes they will reach their destination of magic world and feel satisfied for a few minutes. When people from many lands participate in the game it becomes challenging. Blue whale is one of the games that draws the player to slavery and to do dangerous things.

If the parent is busy, and the child is left without supervision the child is drawn to play such games. Some parents are foolish enough to boast their children's activities and how they have won in computer games. They praise them for their intelligence.

Effects on Body and Mind

Disorders of mind and body show up when game addiction becomes uncontrollable. The children who have attention deficit hyperactivity disorder are prone to be addicts of games online (ADHD). Mostly they are restless. They would be thinking of which game to play next. They will not have much of social life. If you take away the mobile from them they will become violent. Their  studies will suffer. The ADHD affected children and those who have anxiety disorder have the tendency to play more games. Youth who are addicted to games will turn out to be lazy and uninterested in work or hobbies. They may lose their jobs. The married people may even end up in divorce. Some may use all their money to play games and resort to criminal activities to get more money. Some may get access to high speed internet through gigabit files. This results in draining of money. There are some people who use diapers constantly so that they can play games for hours together. There is news of a person who was paralyzed from waist downwards by sitting before the computer playing games.

Parents will fail to control the children after a stage. They will not agree to be taken to the hospital. Even if they are brought before a psychiatrist they will not cooperate with the doctor. They may react violently, those who spend long hours at night become the prey of insomnia. Irregular eating habits will lead to stomach disorders. Habit of eating junk food while playing will worsen the problem. The constant use of mobiles leads to migraine and vision disorder.

They are prone to computer vision syndrome and carpel tunnel syndrome. The addicts may even become careless about basic hygiene.

Treatment

When you realize the addiction of the child go to the doctor without him/her and take his advice. De-addiction treatment for internet addiction in the same as for other addictions. Bupropion the medicine for treating smoking addiction has proved helpful for internet addiction also. This will lessen the strong urge. The treatment for depression helps to curb the negative effects of internet addiction. There are medicines for acute anxiety. Those who have ADHD and have become intolerable must be given medical treatment. They may be admitted to the hospital and treated under specialist doctor and nurses. Access to mobiles should be prohibited. With the treatment of cognitive behavioural therapy behavioural patterns can be improved. Alcoholic Anonymous uses a 12 point therapy for the patients. The same can be given to internet addiction.

Preventive Measures

Children should be aware of the bad effects of mobile addiction. They must be kept away from mobiles, laptop. The parents must be aware of the games they play with these gadgets. Do not allow them to play games daily. There should be a time chart for using the mobile. The computer and the mobile should be switched off two hours before time for sleep. Net service should be cut at a pre-planned time. Arrange the computer at an open place where everyone can watch the user. Cooperate with science clubs and parent-teacher association to be aware of new malicious practices. Make children aware of these pit falls. The in-built games on the mobile may be removed to curb the tendency to play games.

Chapter 14

What is Blue Whale Game?

Blue whale is a mind manipulating game. It cannot be downloaded like other games on the mobile or computer. The access to the game is only through the internet. The link to the game is available only through the secret groups on the social web site.

The Russian psychology student Philip Budeekin programmed the game at 21 years of age. When found guilty, he said, 'I am trying to eliminate the bad elements of the society'. Those who commit suicide are the biological waste materials that should be discarded. The society needs purification'.

Philip justifies his practices and thinks that all who commit suicide are anti-social and a type of depressed people. The game is supposed to be launched in 2014. Russian Publication Novaya Gazette reports about 130 suicides which were committed due to the blue whale game.

Luliya Kontandinova of 15 years of age, and Veronica Volkava of 16 years of age gave information about the blue whale game before committing suicide. Only then the world came to know of the dangerous game.

This game is banned everywhere now. But it comes out in many other names. Experts say it is difficult to remove it completely.

The problem of blue whale is not confined to the foreign countries. Even in our small state Kerala, the use of smart phones and internet is above the percentage of national use. The number of mobiles exceed the population.

The blue whale games last for 50 days. Every day the admin will entrust the player with a particular task. Making cuts on the body, afflicting pain on oneself or visiting the symmetry at midnight, seeing a horror film are some of the tasks given. After fulfilling the tasks the player must send pictures as proof. Then the admin will praise and permit the player to continue. The player is put in a position where he cannot turn back. The fiftieth task is to make a record of your own suicide. At this stage the player would have turned into a complete slave.

The blue whale game makes an addict of the player. He/she obeys the commands of someone on the internet and commit harmful practices to own self. The Russian website 'we contact' has wide spread popularity in Russia. A secret group A57 began the blue whale game in 2013. Now it has been blocked on the internet. Only those who have secret access can get into the site. Budick says that the biological waste created by people who have a tendency to commit suicide must be removed from the face of the world.

When a player starts the blue whale game he/she is asked to make the picture of the blue whale inside the hand by the use of a sharp blade. We may wonder why people are so foolish to hurt themselves. But the fact is hundreds of teenagers are ready to obey the commands of the admin of blue whale.

Are these teenagers 'biological waste? To get an answer we must analyse the true nature of the players. Are they all likely to commit suicide? No, but the main addicts are teenagers between 12 and 20. They are at an age of thoughtless conclusions and an urge to become heroes in the eyes of the world. Blue whale claims success on these supple minds.

There are some special features in children who are likely to become prey to the blue whale.

1. Lonely children: They talk less, and have no bonds with the society. Their world consists only of the internet and computer. Their curiosity will lead them to secret groups and dangerous games. Their curiosity may lead to find the game by chance. Then they go on to become heroes in the eyes of the world with an inner urge.
2. The children who have inferiority complex will also become addicts because they get praise and recognition when they complete difficult tasks. They appreciate this more because they lack the recognition of the society.
3. The people who spend a lot of time on social web sites also falls prey to dangerous games.
4. Some people are always on the lookout for new experiences and heroic deeds. They enjoy the rush of adrenalin while completing an impossible task. A toxic effect is felt when dangerous deeds are fulfilled. They try to initiate the heroes of races, gambling by completing dangerous tasks.
5. Some children feel neglected. The parents have little time to spend on them because of job pressure or other reasons. The children are not given proper instructions of what to do and what not to do. They cannot distinguish right from the wrong. They fall prey to the blue whale because the admin gives them more attention and praise. They get recognition from other players also.
6. Some children suffer from personality disorders. They are emotionally unstable. They are disinterested in the real world. Even if they make friends, it will only last for a few days. They are always unstable. They are quick to flare up in rage and

shout and scream at any time. Suicidal tendency hides behind such personalities. Even small problems lead them to suicide. Blue whale game brings out this tendency and leads them to a sad end.

Chapter 15

Selfitis – The New Disease of the New Age

Do you remember the words of the celebrity who wrote in your autograph? How you persuaded the celebrity and managed to get the lines written? Now it may lie somewhere forgotten among the books in your shelf.

Nowadays autograph fashion has vanished. It has been replaced by selfies. You run after celebrities to take photos on the mobile with posing near them. The celebrities in the field of cinema and politics are very popular. The people who see the photos will appreciate you.

The word selfie was coined only in 2014 and accepted by the English dictionary. The Oxford Dictionary defines selfie as a picture of oneself intended to be posted on by making use of the mobile camera or web camera. When there are many people the picture is known as 'groupfie'.

The idea of photography has completely changed by the onset of mobiles. In past years taking the photo was a rare affair. A photographer was needed. Some money had to be spent. There would be delay in processing the photo.

Now the photos can be taken at anytime and anywhere. Even official events and marriage ceremonies can be taken without the aid of a professional photographer. These are achievements of modern technology.

There are two sides of the coin. The opposite effects created by modern technology become unbearable. They have a negative power on people and society. We have proven ourselves best in turning modern technology in a negative way.

Making use of the positive sides and quickly moving towards growth should be our aim in life. But now we use technology for getting small pleasures and reveal its reverse nature. The fore thought is found lacking. The same happens in the case of selfies.

Selfitis is a disease when there is no control of taking selfies. The continuous posting of selfies in the social media is a compulsive disorder.

The American Psychiatric Association has officially declared selfitis as a mental disease. Do not find fault with the psychiatrist and say that it is only a doctor's whim. Selfitis is one of the diseases which needs serious attention.

There is no harm if you are taking selfies once in a while. But if you are indulging in taking photos of yourself at all times and show eagerness to post them on the social media continuously people will feel that there is something wrong with you. Some take pictures in nude. Some pose near accident victims without trying to help them. Their aim is to attract attention of strangers and collect as many likes as possible. Society regards them as mental patients.

The tendency of selfitis and seeking of likes have given rise for the production of beauty creams and lotions by branded companies. Costly mobiles with cameras which can take clear photos have flooded the market. The market persuades us to believe that the colour, tone and shine of the skin and beauty of the features and body is more important than anything.

Some people have committed suicide because they did not get enough likes for their posts on the social media. The accident victims are left helpless when selfie seekers rush to the site of accident. Such incidents lead the psychiatrists to name selfitis as a disease.

American Psychiatric Association see selfitis as a behavioual addiction disorder. Addiction puts pressure on the mind and becomes worse with time. Withdrawal symptoms are felt when trying to move away from addiction.

There are three types of selfitis

1. Borderline selfitis

When one indulges in selfies for three times a day and post them on social media you have borderline selfitis.

2. Acute selfitis

Acute selfitis make the person take selfies more than three times and put them on the social media.

3. Chronic selfitis

Chronic selfitis is the urge to take more than six selfies a day and constantly post them on the social media. Acute selfitis and chronic selfitis need the guidance and attention of a psychiatrist.

The Thyagaraja University and the Nottingham University have joined hands to develop a scale for selfitis. They did experiments on 200 people and made a selfitis behavioual scale to assess the intensity of the disease. The scale proved effective when a survey was conducted on 400 people. The experiments were conducted in India because face book lovers and cyber borne suicides were the highest in this country.

People who have a healthy mind will not get addicted to selfies. Those who suffer from other maladies turn into selfie addicts. Some may suffer from body dysmorphic disorder.

The mental disorder of body dysmorphophobia leads the patient to feel some parts of the body are not adequate and the leads to depression. They will be persuaded to look at these parts by taking selfies and check whether they are abnormal.

Those who have obsessive compulsive disorder may also suffer from selfitis. They will constantly get into the interested site and neglect all other activities. Selfie addicts enjoy their beauty and have pride in their figure. They worry about small blemishes and changes to their beauty and complexion.

Attention seekers are also selfie addicts. When they see new things or share news with others they feel great. Some suffer from inferiority complex and find solace when they get recognition and likes for their selfies.

Narcissism is a state when a person is proud of himself. Narcissist personalities feel superior to others. They will try to declare their supremacy and try to prove that they are always right. The number of likes they get for their selfies will make them feel great and to think that they are always right. It will make them proud and indulge in more selfies.

Those who challenge the rules of the society and neglect to follow them are psychopaths. This personality leads them to selfitis. They lack sincerity, positive thinking and feel guilty. They will not allow themselves to be guided by the inner voice which points out the right from the wrong. They become the prey of selfitis and feel depressed when enough recognition is not got from the society. Their ego will suffer and sometimes lead to suicide.

The people who have mental disorder go on taking selfies to appease an inner urge. They become addicts in the course of time.

Keralites are the largest number of people who make use of modern technology among Indians. The most modern gadgets are seen in the homes. Even little children know to operate them. The number of people who take selfies while driving is on the increase. The addicts pose on top of a rock or cruise or before an oncoming train to show their heroism. When a person has fallen from a building and is lying on the ground bleeding, people rush to the scene to make gruesome selfies. When a poor mental patient is attacked by a crowd instead of helping the victim the selfie addicts are busy taking pictures. The society should take action to change this attitude.

Bayers Pharmaceuticals have come up with a drug which prevents the selfie urge with the consent of United States Food and Drug Administration (FDA). Bayers claim that the drug is anti-selfie and will ooze the urge from the mind. They suggest a green pill for borderline cases, a blue pill for acute illness and a red pill for chronic selfitis.

Details of their pill are unknown. Usually permit is given by FDA only after some years. This medicine put up by Bayers was given permission within 4 months. These facts should be taken into consideration.

New Stanford has made a bread toaster and put it on the market. In every slice of bread your face will be imprinted and by toasting it and consuming it you may get out of your selfie addiction. They claim that the ego will be satisfied. The fact that behavioural addiction cannot be cured easily with medicines should be taken into consideration.

Selfitis has reached a stage when medicine has to be used. The big pharmaceutical companies are taking advantage of this and are filling the market with costly drugs. It also shows that many foreign countries have recognized selfitis as a disease.

The experts in psychiatry prefer to avoid medicines and try behavioural therapy for selfitis. Anti-addiction medicines which are effective for other addictions may be useful in the treatment of selfitis. Behavioural changes must be treated in other ways.

The reason for making a person an addict must be found out. If the mental state is changed, then the medicines for addiction may prove effective.

Habits are displaced by new healthy habits by behavioural therapy. The interest in selfie can be turned into a group selfie interest. This changes the selfish interest in oneself. It will also change inferiority complex and the impact of ego.

Narcissist personality, psychopathy, inferiority complex and craving for the attention of others can be changed through behavioural therapy. But serious narcissism and psychopathy are very difficult to be cured.

Cognitive behavioural therapy is a means to change the pattern of thinking. Thus misunderstandings in the thought process are changed and a new approach to life is put in. This is a cure for depression and body dysmorphic disorder and OCD which lead to selfitis. The medicines given for depression are found helpful in the treatment of selfitis also.

What can the society do to prevent selfitis? The society should discourage pictures which are taken while driving or in dangerous places or before a running train. It should not encourage practices like taking selfies in accident sites. These pictures should not be published. The friends, parents and others have a role to play to discourage selfies. The selfie mania should be put on end to.

The celebrities have a responsibility to discourage taking pictures with them. The media should not give much importance to selfies. People should realize that they are missing the beauty of nature and surroundings by taking selfies all the time.

Selfies have become part of our life and culture. When it becomes a waste of time or dangerous or leads a person to showmanship it becomes a disease. One should be aware that selfitis is a disease and a mental disorder.

Selfitis Behavioural Scale

Answer these questions truthfully and mark the points

1. Taking a selfie makes me enjoy my surroundings
2. I take it as a healthy competition when sharing selfies with friends and co-workers
3. I get a lot of enjoyment and encouragement on the social media for my selfies
4. I resort to selfies to lessen mental pressure
5. I feel self-confident when i take a good selfie
6. When I share selfies with friends I feel more recognized
7. I feel comfortable and find it easy to adjust to my surroundings
8. I feel great when I post selfies in different poses
9. I feel recognized by the society when I get more likes for my selfies
10. When I take selfies I feel elated
11. I have become famous among my friends because of my selfies
12. Selfies give me hope
13. Selfies help me to remember and enjoy an occasion
14. I post more and more selfies to get more likes and comments
15. I expect more recognition from my friends
16. My mental state changes suddenly when I take a selfie
17. I enjoy my selfies in secret. It gives me confidence.
18. If I do not post a selfie I feel that I have become an outsider.
19. I make selfies to remember the occasion in future
20. I regularly use editing software to enhance the beauty of selfies.

(On a scale beginning from one to five which will determine the intensity of your love for selfies, calculate points. If your points add up to more than 50 you are an addict. The minimum points are 20 and maximum 100).

Chapter 16

The Use of Mobiles and Health Problems

Health Problems and Family Social Problems

The constant use of mobile phones causes long time effects on health and mental state. The researchers in London say the mobiles are like time bombs. The research centre came to a conclusion after studying 200 cases. Long term constant use of mobiles may lead to tumors in the brain like glioma. The symptoms take years to develop so the effect cannot be predicted at this time. The mobiles came to be widely used in 1990. Long term conclusions are to be awaited.

Cancer

There are many results that point out mobiles may cause health problems or even cancer. They send non-ionizing waves which are a type of radio waves. They are not as harmful as ionized waves. But the cells near the ear may have effect with the constant use of mobiles. The National Agency for Cancer Research has included non-ionizing radiation waves among the reasons for cancer. The electrical impulses raise the temperature and creates a biological change in the body. 50 minutes of constant mobile use increases glucose metabolism in the nearby brain cells. 20 minutes of use raise the temperature in the cells of ear by one degree. The World Health Organization made a study of the impact on brain cells by the constant use of mobiles in 50 countries and came to a conclusion that damage to cells cannot be written off.

Glioma and Meningitis may affect the brain. The cells near the ear may have a dangerous effect. A study conducted in Sweden also says that brain cell damage is possible by the constant use of mobiles. Those who are badly affected are children below the age of 20.

How to lessen radiation?

The safety depends on the lessening of radiation. Energy emitted by the mobile is calculated to find how much damage it may cause. In the place where there is good mobile range minimum energy is emitted. The radiation effects are low.

The radiation effect from the mobile is known as Specific Absorption Rate Value (SAR). The calculation can be done by observing the radiation absorbed by the body when using the mobile. If SAR is less radiation emitted is less. In America and Europe only mobiles which have an SAR below 2 watts/kilogram are allowed to be marketed.

The use of mobiles by children must be restricted. Their skull is thin and brain is small. So the radiation rays can get inside the brain easily. We cannot predict the ill effects in future at this stage.

Mobile Tower and Health Problems

The mobile phones are connected to base stations by mobile tower. The mobile companies in their competition to enhance coverage have constructed a number of mobile towers. The statistics say that there are 4.3 lacs of mobile towers in the country. Mobile towers constantly send electromagnetic radiation to surroundings.

Indian Council of Medical Research has found out that people who stay near a mobile tower will experience tiredness and insomnia. The effect is more who reside within 50-300 meters of the tower. The radiation has effect not only on humans but also on animals and birds. The milk production becomes less in the livestock. The radiation blocks the free air ways of the birds. The bees may find it difficult to go back to their hives because of radiation waves.

Other Problems

The waves emitted by the mobile towers may cause head ache, insomnia, anxiety, depression and lack of appetite in some people. A study done in Tel Aviv University of Israel says that people who live within the radius of 350 meters of the tower are prone to lung cancer and cancer of the kidneys. Women are more affected by the waves. In men the movement of sperms may become dull.

Pain in the Fingers

Sending messages continuously will experience pain and numbness in fingers. The messaging should be stopped immediately. Some other fingers may be tried to send messages. Those who type for long hours on the key board can also experience similar pain.

Carpel Tunnel Syndrome

The elbow gets strained by the constant use of mobiles. The elbow may experience numbness or pricking pain. The ulnar nerve becomes pressurized because of the bending of elbow.

Hearing Disability

Studies say that frequent users are prone to loss of hearing. The ENT Branch of Chandigarh Post Graduate Institute of Medical Sciences have proven this by their study. The study says that constant use of mobiles lead to ear ache.

Infertility

The study conducted in Cleveland and Hungary has found out that mobiles effect sperm count, mobility of sperms and their quality. The mobiles kept in pant pockets and belt pouches have an effect on genital organs.

Pregnant Women

Women should be careful in their use of mobiles when pregnant. A study done at Los Angeles University of Canada says that radiation may affect the unborn child. They may have a tendency to be hyperactive. The unborn child and a child till the age of 7 are more effected by radiation.

Serious health disorders are seen in the people who use mobiles constantly. The bad effects are more on children. Their power of vision may be affected. They may suffer from ear ache. Mobile use for long hours may affect blood pressure and diminish the production of red blood cells. Bangladesh has recently prohibited the use of mobiles by pregnant women and children.

Slavery to the Mobile

The slaves of the mobile watch the mobiles for messages often. They also send messages constantly. Every hour they may check the mobile phone and play games. If they forget to take the mobile they feel anxious and are tensed. They may have breathing difficulties and mental break down if they are kept away from their mobiles.

Road Accidents

Sending messages while driving or reading messages result in road accidents. We can also see any number of drivers with bent necks talking on the phone. Using mobile while driving is as dangerous, as drunken driving. Both are punishable by law. National Safety Council says that more than one among four road accidents are due to mobile usage. Japan, Brazil, Australia, Italy, France, Singapore, Ireland and Britain have declared using mobiles while driving as unlawful. United States of America does not find it unlawful when hand sets are used while driving.

Reasons for accidents

1. The driver's attention distracts when talking on the phone.
2. It makes the phone a foe than a friend. It breaks up relationships and social ties
3. Land phones were the property of everyone at home. Now the mobiles are personal property and secret devices. Each member owns a mobile of his/her own. Rules of privacy and secrecy are set by each one. Selfishness increases.
4. This use of mobile as a secret device creates a secret world of one's own where one can spend time with selfish pleasure.
5. The mobile addicts do not communicate with family members. They forget to connect with far away friends and relatives. Mobiles have also re-defined the social life.

Change in Behaviour

The rhythm of life is affected and the addict becomes more unruly in habits. The addict goes to sleep late. He develops a tendency for pornography. They tend to have secret relationships and love affairs which absorb their time. They drift away from studies. They have sleep disorders and irregular food habits. Their health is affected. Many parents bring their children to the psychiatrists. These children may always talk in a low voice over the phone or see clippings all the time.

Unhealthy Relationships

Secret relationships from the past are revived or new relationships are made through the use of mobiles. The relationship changes according to the personality of the mobile user. Indecent practices like sexual talk and relationships will arise. These unhealthy relationships will develop through missed calls, anonymous calls and messages. Anonymous calls are always a sign of danger.

Mobile Camera

Today the mobile is more than a communication device. It has become a toy, or music player or camera or computer. The camera is the most misused part of the mobile. The criminals in the society take secret photos and try to blackmail the victims. This has become a social evil.

Awareness, attentiveness and severe punishment are the only means to prevent these crimes which are increasing in the modern society.

Forgetfulness and Laziness

Before the onset of mobile phones we used to know at least 20 phone numbers of our relatives and friends by heart. We also remembered STD codes of different places. Now even our own phone number is difficult to remember. When the mobile is lost you are helpless without even a single phone number of your friends or relatives in your memory. As the mobile provides a calculator even small purchases have become difficult without the mobile.

Mobiles are helpful in many ways. But these devices make the brain lazy and useless. Sharpness of memory is lost. Intelligence becomes lost. The power to differentiate right from the wrong is lost. This is not good for the society.

Chapter 17

Treatment

The treatment for mobile addiction begins with the restriction of time for using them. At first the addicts will resist but after a time they will compromise and find joy in the peacefulness they get when away from mobiles. Sometimes even the health problems may begin to vanish. They will view the past with contempt.

Therapists have an important role to play in the treatment of mobile phone addiction. When the addiction leads to inefficiency in jobs or depression therapy is needed. Therapists will treat the addiction as well as probe into other phobias and defects of the patient.

Some ways are suggested to reduce mobile phone addiction

1. Think of the time you spend

Calculate the time you spend on the mobile for talking or sending messages. Note the time in a book, continue this for a week. Then analyse the habit.

2. Try to put an end to the mobile habit. Restrict the messages. At first the use of mobile can be restricted 10%. Then gradually come down from ten hours to nine and so on.

3. Live in the present

People talk on the phone when they have nothing else to do. We can often see people using the mobile while waiting on a queue or bus station.

Keep the mobile switched off when talking with friends. Keeping it away fom you will resist your temptation to look at it all times. Try to enjoy living in the present.

4. Unnecessary connections

Do you feel it is most necessary to have a mobile for communicating with each other? They are not necessary to cultivate healthy friendships. Love relationships do not grow through words. Give it time to grow in a natural way.

5. The feeling of importance

As an executive you think that you have to attend to all communications through the mobile. You feel it is important to attend to calls even when driving. The mobile is not switched off because you think that some urgency might arise even at midnight. The checking of missed calls and messages can be done in the morning. Recognize that you are not that important to be on the line at all times.

6. Technology is for our benefit

If modern technology adds pressure, anxiety and economic problems in your life move away from them. Technology is something that should make life easy. Keep a fixed time to check messages.

Mobile phone addiction should not destroy your life, job and relationships. Even after practicing these steps you fail in your attempts you should seek the help of a psychiatrist.

Mobile Phone Etiquette

There are some common rules which we should follow in life. The main point is to give respect to others. The same rule is applicable in the use of mobiles. The first and foremost rule is to keep away from the mobile while driving. High concentration of eyes, ears and hands are needed for driving. Even a small fault can end in a big damage to life and property.

It is also against law. Talking on the mobile while crossing a road/rail track is very dangerous. Many accidents can be avoided this way.

Food habits

Dining table is a place when the family should eat together and share their experiences. Do not use the precious time on the mobile. Do eating without the mobile and enjoy the closeness of your loved ones.

Be courteous in conversations

Do not talk in a loud voice when using the mobile and disturb people nearby. In public places like restaurants, cinema halls go out if you have to attend a phone call. While travelling by bus or train speak in a soft voice and use head phones to listen to music. Using the speaker device in public places is also against etiquette.

Maintain silence

Keep the phone switched off when you go to a place of death and others are mourning. You are being unrespectful to the dead person and his family by using the mobile. When participating in a meeting switch off the mobile. Mobiles are not allowed inside class rooms.

Do not disturb sleep

Public places like hotels, dormitories, night trains, and sleep coaches are places where we should not disturb others while sleeping. Loud conversation on the phone will keep others awake and irritated. Respect their privacy.

Observe the laws

In all places where use of mobiles are prohibited like courts, shrines, petrol pumps, examination rooms of doctors, and lifts switch off the mobile or keep it in a silent mode. Do the same in hospitals and ICU when mobile phones may cause harm by emitting waves.

Prohibition of Photos

Taking photos of another person without permission is unlawful. The accident victims are now open to mobile addicts creating disturbance to medical helpers. Taking a photo of the scene is more important to them than helping the victim.

Use head phones for music

It is not proper to disturb others by playing loud music over the phone. Make use of the head phone if you want to enjoy music in public places.

A decent ring tone

Noisy, mind blowing or pornographic ring tones are against decency.

Messages

Do not send messages while attending a meeting or conversing with a person. This shows disrespect.

Wrong numbers

Check the number before dialing. This will avoid contacting a wrong number. Say sorry to the person if you have dialed a wrong number. If you happen to receive a wrong number call, dismiss it with decency. Do not show anger.

SMS, what Sapp messages

Sending indecent messages or pictures through what Sapp is against law. Do not respond to messages from an unknown source. Make sure the information is reliable before sending it to others.

The mobile services

Make use of the reliable service companies.

The alertness of the telecom authority

India has a population which is reaching the maximum use of mobiles in the world. The telecommunication department is instructing half of the population to be alert and resort to healthy practices. They warn us about the dangerous electromagnetic waves emitted by mobile phones. The damage caused by radiation is reminded. These are the basic rules.

1. In advertisements do not use the pictures of children and pregnant women using the mobiles.
2. Pregnant women, heart patients and children should limit the use of mobiles.
3. Children below 18 should not be given a mobile of their own.
4. The radiation frequency energy may affect the brain cells and make them inactive.
5. Instead of keeping the mobile near your ear, make use of earphones.

Misuse of mobiles

The IT Act declare the misuse of mobiles as a punishable offence.

Taking nude pictures and sending them to others is an offence. Section 67 of the IT Act punish the people who spread pornographic messages and conversations. A fine of 5 lakhs is imposed. If repeated the punishment is raised to 10 lakhs fine and 5 years imprisonment.

Taking pictures of private parts and publishing them in punishable by IT Act 66 for three years imprisonment and a fine of two lakhs. For spreading pornographic videos punishment is declared in section 67A as five years imprisonment and a fine of 10 lakhs. If repeated the term of imprisonment will become 7 years. Downloading sexual related pictures of children or videos and publishing them are punishable by section 67B.

A person who indulges in such practices is punished for imprisonment for 5 years and fine of 10 lakhs. If the offence is repeated the fine is 10 lakhs and imprisonment is for 7 years. These are non-bailable cases. Some daily messages which may turn indecent and unlawful should be deleted immediately. The victim of any such crime should report at the nearest police station. They will turn over the case to the cyber cell.

www.ingramcontent.com/pod-product-compliance
Lightning Source LLC
Chambersburg PA
CBHW030847180526
45163CB00004B/1474